AF270427

ACCIDENTAL
SCIENCE DISCOVERIES

CORN FLAKES

Kenny Abdo

Fly!
An Imprint of Abdo Zoom
abdobooks.com

abdobooks.com

Published by Abdo Zoom, a division of ABDO, P.O. Box 398166, Minneapolis, Minnesota 55439. Copyright © 2024 by Abdo Consulting Group, Inc. International copyrights reserved in all countries. No part of this book may be reproduced in any form without written permission from the publisher. Fly!™ is a trademark and logo of Abdo Zoom.

Printed in the United States of America, North Mankato, Minnesota.
102023
012024

Photo Credits: Alamy, Bridgeman Images, Getty Images, Shutterstock, ©Grosscha p.15/ CC BY-SA 3.0
Production Contributors: Kenny Abdo, Jennie Forsberg, Grace Hansen
Design Contributors: Candice Keimig, Neil Klinepier, Colleen McLaren

Library of Congress Control Number: 2023938009

Publisher's Cataloging-in-Publication Data

Names: Abdo, Kenny, author.
Title: Corn flakes / by Kenny Abdo
Description: Minneapolis, Minnesota : Abdo Zoom, 2024 | Series: Accidental science discoveries | Includes online resources and index.
Identifiers: ISBN 9781098284091 (lib. bdg.) | ISBN 9781098284817 (eBook) | ISBN 9781098285173 (Read-to-Me eBook)
Subjects: LCSH: Cereal products--Juvenile literature. | Kellogg Toasted Corn Flake Company--Juvenile literature. | Serendipity in science--Juvenile literature. | Inventions--Juvenile literature. | Discoveries in science--Juvenile literature.
Classification: DDC 500--dc23

TABLE OF CONTENTS

CORN FLAKES

Corn Flakes may have been discovered by accident, but that error has provided people with a healthy breakfast for more than a century!

THE ACCIDENT

John Harvey Kellogg and his brother Will Keith ran a health **clinic** in the late 1800s. They believed that a simple **vegetarian** diet was the key to good health.

The Kelloggs experimented with different grains in 1894. They accidentally left a pot of boiled wheat out overnight. The next day, they discovered that it had turned into thin, crispy flakes.

THE DISCOVERY

The Kelloggs decided to try the same process with corn. In 1895 they created the first batch of Corn Flakes. By 1896, they **patented** and produced Corn Flakes **commercially**! Kellogg's became a company in 1906.

Kellogg's
CORN
FLAKES

The cereal still tasted a bit bland. So,
Will Keith added sugar to the recipe.
The popularity of Corn Flakes then
skyrocketed.

By the early 1900s, many households had boxes of the cereal in their cupboards!

Kellogg's produced millions of pounds of Corn Flakes during **World War II**. It was a fast and easy way to keep the troops fed. The company also released a new product called Pep. It was a vitamin-rich cereal that was sold as a wartime food.

Kellogg's
PEP
IN ITS OWN
Kel-BOWL-Pac
The Box is the Bowl
Kellogg's
RICE
KRISPIES
OVEN POPPED RICE WITH SUGAR,
SALT AND MALT FLAVORING
NET WEIGHT
ONE OUNCE
MADE BY KELLOGG COMPANY, BATTLE CREEK, MICH.

It's Kellogg's Sweet-eatin' carnival time!
Buy any two packets of these cereals and . . .
Kellogg's will send you a BRAND-NEW SHILLING!
Kellogg's
ALL STARS
Kellogg's
ALL STARS
Oat Cereal
Kellogg's
COCO POPS
CHOCOLATE-FLAVOURED TOASTED RICE
Kellogg's
SUGAR RICICLES
NODDY'S favourite breakfast
Kellogg's
SUGAR SMACKS
Kellogg's
SUGAR SMACKS
Kellogg's
FROSTIES
SUGAR FROSTED FLAKES
SUGAR TOASTED for SUPER ENERGY
"Tony"
Tony
Kellogg's are sure you will enjoy these sugar-sweet cereals—they are making this special offer so that you can try them. This is all you have to do:—
1 Buy any two packets of these cereals and cut off the tops.
2 Write your name and address on the back of one of them.
3 Mail the two tops in an envelope to Kellogg's, Dept. S, Stretford, Manchester.
Kellogg's will send you a brand-new shilling!
Offer closes August 31st, 1961, so send for your shilling quickly. Offer limited to one per family.
SUGAR SMACKS
the Yogi Bear cereal. Honey-sweet puffs of wheat.
FROSTIES
crisp, sparkling flakes. Tony the Tiger says, they're SCR-R-R-umptious!
ALL STARS
the new star-shaped oat cereal with a hole in the middle and sugar on top. The Wizard of Oatz brings magic to breakfast!
SUGAR RICICLES
Little Noddy's favourite breakfast. Crunchy sweet, light and crisp.
COCO POPS
chocolate-flavoured puffs of rice. Coco the Monkey's "anytime" cereal.
FUN MASKS FOR YOUR CHILDREN!
There's a "dressing-up" mask on each packet: Yogi Bear and his T.V. friends —all ready to cut out. So let your children choose and change—for the fun of it. And for the goodness and energy too, in Kellogg's pre-sweetened cereals.
Make it a sweet life for your children
BUY Kellogg's SUGAR-SWEET CEREALS TODAY!

Kellogg's continued to **innovate** its cereals through the years. The company remained focused on health and wellness. It introduced more cereals that were low in sugar and high in **fiber**.

Eggo®

Homestyle

NO ARTIFICIAL COLORS OR FLAVORS
9 VITAMINS AND MINERALS

Kellogg's

Eggo

Buttermilk

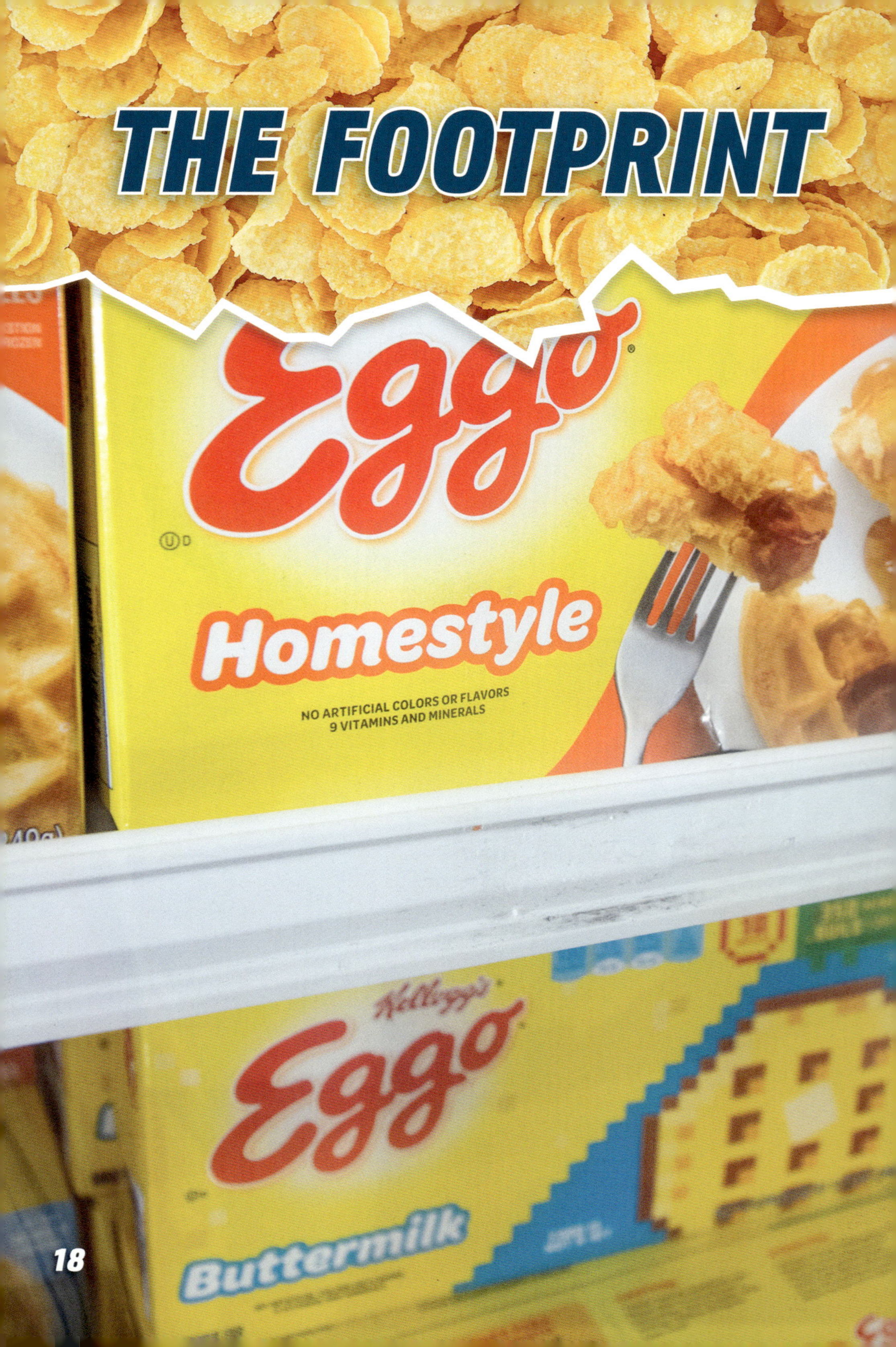

Today, Kellogg's is one of the largest cereal companies in the world. The company also sells products such as Pop-Tarts and Eggo Waffles in over 180 countries!

Corn Flakes remains one of Kellogg's most popular products. It is enjoyed by millions of people around the world. The Kellogg brothers created a food that does not flake out on its customers!

GLOSSARY

clinic – a facility that is focused on the health of patients.

commercial – the mass production of something that is sold to the public.

fiber – a plant material found in food that is not digested by the body but plays an important part in good health.

innovate – to come up with and implement a new idea, method, or device.

patent – the exclusive right granted to a person to make or sell an invention. This right lasts for a certain period of time.

vegetarian – a diet consisting mostly of plant foods and sometimes eggs or dairy products. Vegetarians do not eat meat, poultry, or fish.

World War II – (1939–1945) a war fought in Europe, Asia, and Africa. Great Britain, France, the United States, the Soviet Union, and their allies were on one side. Germany, Italy, Japan, and their allies were on the other side.

ONLINE RESOURCES

Booklinks
NONFICTION NETWORK
FREE! ONLINE NONFICTION RESOURCES

To learn more about Corn Flakes, please visit **abdobooklinks.com** or scan this QR code. These links are routinely monitored and updated to provide the most current information available.

INDEX